THE HYDROGEN ECONOMY

The Fuel Of The Future

Navneet Kumar

ABSTRACT

Between production and use, any commercial product is subject to the following processes: packaging, transportation, storage, and transfer. The same is true for hydrogen in a "Hydrogen Economy". Hydrogen has to be packaged by compression or liquefaction, it has to be transported by surface vehicles or pipelines, it has to be stored and transferred. Generated by electrolysis or chemistry, the fuel gas has to go through these market procedures before it can be used by the customer, even if it is produced locally at filling stations. As there are no environmental or energetic advantages in producing hydrogen from natural gas or other hydrocarbons, we do not consider this option, although hydrogen can be chemically synthesized at a relatively low cost.

In the past, hydrogen production and hydrogen use have been addressed by many, assuming that hydrogen gas is just another gaseous energy carrier and that it can be handled much like natural gas in today's energy economy. With this study, we present an analysis of the energy required to operate a pure hydrogen economy. High-grade electricity from renewable or nuclear sources is needed not only to generate hydrogen but also for all other essential steps of a hydrogen economy. But because of the molecular structure of hydrogen, hydrogen infrastructure is much more energy-intensive than a natural gas economy.

In this study, the energy consumed by each stage is related to the energy content (higher heating value HHV) of the delivered hydrogen itself. The analysis reveals that much more energy is needed to operate a hydrogen economy than is consumed in today's energy economy. Depending on the chosen route the input of electrical energy to make, package, transport, store and transfer hydrogen may easily double the hydrogen energy delivered to the end-user. But precious energy can be saved by packaging hydrogen chemically in a synthetic liquid hydrocarbon like methanol or dimethyl ether DME. We, therefore, suggest modifying the vision of a hydrogen economy by considering not only the closed hydrogen (water) cycle but also the closed carbon (CO_2) cycle. This could create the intellectual platform for the conception of a post-fossil fuel energy economy based on synthetic hydrocarbons. Carbon atoms from biomass, organic waste materials, or recycled carbon dioxide could become the carriers for hydrogen atoms. Furthermore, the energy-consuming electrolysis may be partially replaced by the less energy-intensive chemical transformation of water and carbon to synthetic hydrocarbons. As long as the carbon comes from the biosphere ("biocarbon") the synthetic hydrocarbon economy would be as benign concerning the environment as a pure hydrogen economy. But the use of "geocarbons" from fossil sources should be avoided to uncouple energy use from global warming.

CONTENTS

Title Page
Abstract
Introduction
1. Properties of Hydrogen   1
2. Energy Needs of a Hydrogen Economy   4
3. Production of Hydrogen   6
4. Packaging of Hydrogen   10
5. Delivery of Hydrogen   19
6. Transfer of Hydrogen   28
7. Summary of Results   31
8. Conclusions   40
Books By This Author   41

INTRODUCTION

Hydrogen is a fascinating energy carrier. It can be produced from electricity and water. Its conversion to heat or power is simple and clean. When combusted with oxygen, hydrogen forms water. No pollutants are generated or emitted. The water is returned to nature where it originally came from. But hydrogen, the most common chemical element on the planet, does not exist in nature in its pure form. It has to be separated from chemical compounds, by electrolysis from water or by chemical processes from hydrocarbons or other hydrogen carriers. The electricity for the electrolysis may eventually come from clean renewable sources such as solar radiation, the kinetic energy of wind and water, or geothermal heat. Therefore, hydrogen may become an important link between renewable physical energy and chemical energy carriers.

Hydrogen has fascinated generations of people for centuries including visionary minds like Jules Verne. A "Hydrogen Economy" is projected as the ultimate solution for energy and the environment. Hydrogen societies have been formed for the promotion of this goal by publications, meetings, and exhibitions. But has physics also been properly considered?

Both the production and the use of hydrogen have attracted the highest attention while the practical aspects of a hydrogen economy, Figure 1, are rarely addressed.

Like any other product, hydrogen must be packaged, transported, stored, and transferred to bring it from production to final use. These ordinary market processes require energy. Hydrogen has fascinated generations of people for centuries including visionary minds like Jules Verne. A "Hydrogen Economy" is projected as the ultimate solution for energy and the environment. Hydrogen societies have been formed for the promotion of this goal by publications, meetings, and exhibitions. But has physics also been properly considered?

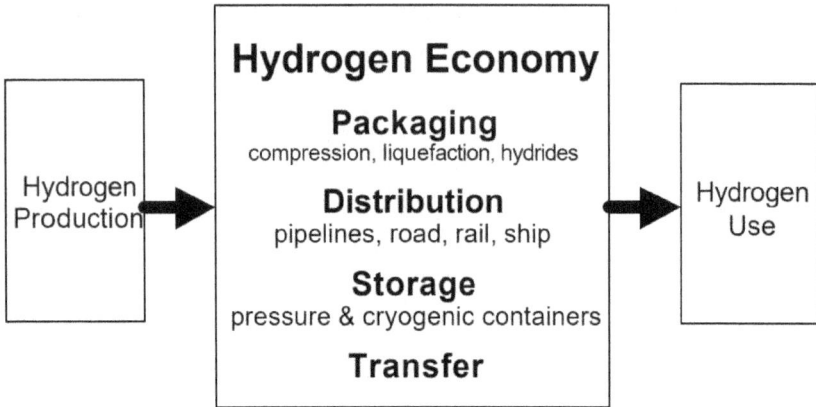

The energy lost in today's energy economy amounts to about 10% of the energy delivered to the customer. We would now like to present rough estimates of the energy required to operate a "Hydrogen Economy".

Without question, technology for a hydrogen economy exists or can be developed. In fact, enormous amounts of hydrogen are generated, handled, transported and used in the chemical industry today. But this hydrogen is a chemical substance, not an energy commodity. Hydrogen production and transportation costs are absorbed in the price of the synthesized chemicals. The cost of hydrogen remains irrelevant as long as the final products find markets. Today, the use of hydrogen is governed by economic arguments and not by energetic considerations.

But if hydrogen is used as an energy carrier, energetic arguments must also be considered [1]. How much high-grade energy is used to make, package, handle, store, or transport hydrogen? The global energy problem cannot be solved in a renewable energy environment if the energy consumed to make and deliver hydrogen is of the same order as the energy content of the delivered fuel. But how much energy is consumed for compression, liquefaction, transportation, storage, and transfer of hydrogen? Will there be only the hydrogen path in the future? We have examined the key market procedures by physical reasoning and conclude that the future energy economy is unlikely to be based on pure hydrogen alone. Hydrogen will certainly be the main link between renewable physical and chemical energy, but most likely it will come to the consumer chemically packaged in the form of one or more synthetic consumer-friendly hydrocarbons.

Preliminary results of our study have already been presented at THE FUEL CELL WORLD conference in July 2002.

1. PROPERTIES OF HYDROGEN

The physical properties of hydrogen are well known. It is the smallest of all atoms. Consequently, hydrogen is the lightest gas, about 8 times lighter than methane (representing natural gas). The gravimetric higher heating value "HHV" of fuel gas is of little relevance for practical applications. In general, the volume available for fuel tanks is limited, not only in automotive applications. Also, the diameter of pipelines cannot be increased at will. Therefore, for most practical assessments it is more meaningful to refer the energy content of fuel gases to a reference volume. Also, it is proper to use the higher heating value HHV (heat of formation) for this energy analysis, because it reflects the true energy content of the fuel based on the energy conservation principle (1st Law of Thermodynamics). By contrast, the lower heating value LLV is a technical standard created in the 19th century by boiler engineers confronted with problems of corrosion in the chimneys of coal-fired furnaces caused by condensation of sulfuric acid and other 5 aggressive substances. Since the production of hydrogen is governed by the heat of formation or the higher heating value, its use should also be related to its HHV energy content. The following volumetric higher heating values for hydrogen and methane at 1 bar and 25 °C will be used in this study.

	Dimensions	Hydrogen	Methane
Density at NTP	kg/m^3	0.09	0.72
Gravimetric HHV	MJ/kg	142.0	55.6
Volumetric HHV	MJ/m^3	12.7	40.0

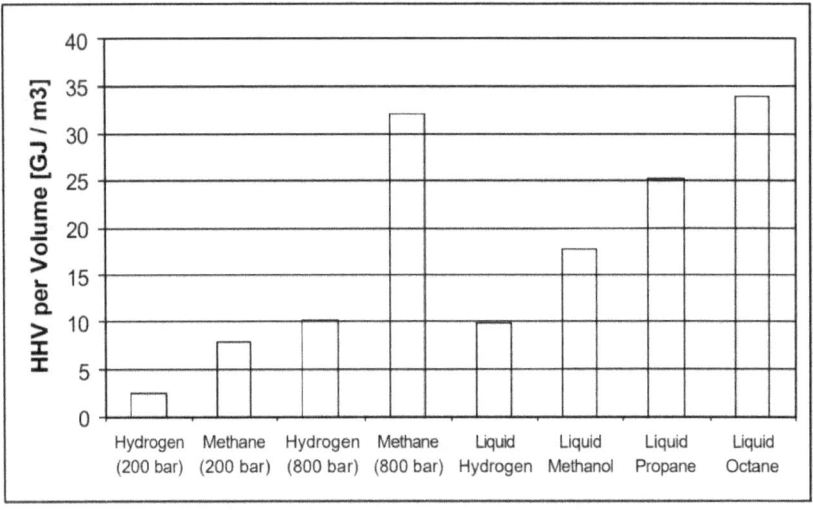

Figure 2   Volumetric HHV energy density of different fuels

figure 2 shows the volumetric HHV energy densities of different energy carrier options. At any pressure, hydrogen gas carries less energy per volume than methane (representing natural gas), methanol, propane, or octane (representing gasoline). At 800 bar pressure, gaseous hydrogen reaches the volumetric energy density of liquid hydrogen. But at any pressure, the volumetric energy density of methane gas exceeds that of hydrogen gas by a factor of 3.2 (neglecting non-ideal gas effects). The common liquid energy carriers like methanol, propane, and octane (gasoline) surpass liquid hydrogen by factors 1.8 to 3.4, respectively. But at 800 bar or in the liquid state hydrogen must be contained in hi-tech pressure tanks or cryogenic containers, while the liquid fuels are kept under atmospheric conditions in unsophisticated containers.

2. ENERGY NEEDS OF A HYDROGEN ECONOMY

Hydrogen is a synthetic energy carrier. It carries energy generated by some other processes. Electrical energy is transferred to hydrogen by the electrolysis of water. But high-grade electrical energy is used not only to produce hydrogen but also to compress, liquefy, transport, transfer or store the medium. In most cases, the electrical energy could be distributed directly to the end-user. For all stationary applications hydrogen competes with grid electricity. Furthermore, liquid synthetic hydrocarbons could also serve as the general energy carrier of the future. Carbon from biomass or CO_2 captured from flue gases could become the carrier for hydrogen atoms generated with electrical energy from renewable or nuclear sources. There are environmentally benign alternatives to hydrogen.

Certainly, the cost of hydrogen should be as low as possible. But the hydrogen economy can establish itself only if it makes sense energetically. Otherwise, better solutions will conquer the market. Also, infrastructures exist for almost any synthetic liquid hydrocarbon, while hydrogen requires a new distribution network. The transition to a pure hydrogen economy will affect the entire energy supply and distribution system. Therefore, all aspects of a hydrogen economy should be discussed before investments are made.

The fundamental question: "How much energy is needed to operate a hydrogen economy?" will be analyzed in detail. We consider the key elements of a hydrogen economy like production, packaging, transport, storage, and transfer of pure hydrogen and relate the energy consumed for these functions to the energy content of the delivered hydrogen. Our analysis is based on physics and verified by numbers obtained from the hydrogen industry. Throughout the study, only representative technical solutions will be considered.

3. PRODUCTION OF HYDROGEN

3.1 Electrolysis

Hydrogen does not exist in nature in its pure state but has to be produced from sources like water and natural gas. The synthesis of hydrogen requires energy. Ideally, the energy input equals the energy content of the synthetic gas. Hydrogen production by any process, e.g. electrolysis, reforming, or else, is a process of energy transformation. Electrical energy or chemical energy of hydrocarbons is transferred to the chemical energy of hydrogen. Unfortunately, the process of hydrogen production is always associated with energy losses.

Making hydrogen from water by electrolysis is one of the worst energy-intensive ways to produce fuel. It is a clean process as long as the electricity comes from a clean source. But electrolysis is associated with losses. Electrolysis is the reversal of the hydrogen oxidation reaction the standard potential of which is about 1.23 Volts at NPT conditions. But electrolyzers need higher voltage to separate water into hydrogen and oxygen. The overpotential is needed to overcome polarization and ohmic losses caused by electric current flow under operational conditions.

The electrolyzer and fuel cell characteristics are schematically shown in Figure 3. Under open-circuit conditions, the electrochemical potential is 1.23 Volts at 20°C.

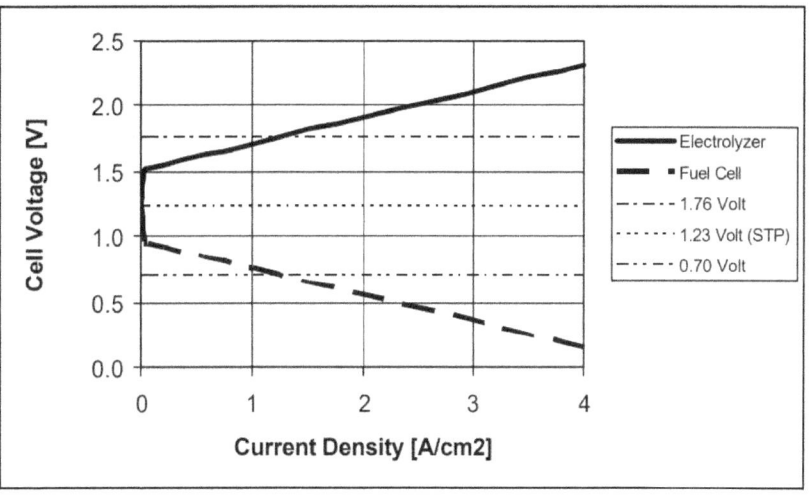

Figure 3   Voltage-current characteristics of electrolyzer and fuel cell.

Assuming that the same electrolyte and catalysts are used, the polarization losses are typically 0.28 Volt for solid polymer or alkaline systems. The apparent open-circuit voltages thus become 0.95 and 1.51 Volt for fuel cell and electrolyzer, respectively. For both, we assume an area-specific resistance of 0.2 Wcm2 and construct the characteristics for a low-temperature fuel cell (dashed line) and a corresponding electrolyzer (solid line).

Fuel cells are normally operated at 0.7 Volt to optimize the system efficiency. We assume the same optimization requirements also hold for an electrolyzer. In this case, the corresponding voltage of operation is 1.76 Volts as indicated by the dash-dot lines in Figure 3.

The standard potential of 1.23 Volts corresponds to the higher heating value HHV of hydrogen. Consequently, the over-potential is a measure of the electrical losses of the functioning electrolyzer. The losses depend on the current density or the hydrogen production rate. As shown in Figure 4, at 1.76 Volt 1.43 energy units must be supplied for every HHV energy unit contained in the liberated hydrogen. At higher hydrogen production rates (higher current densities) this number increases further.

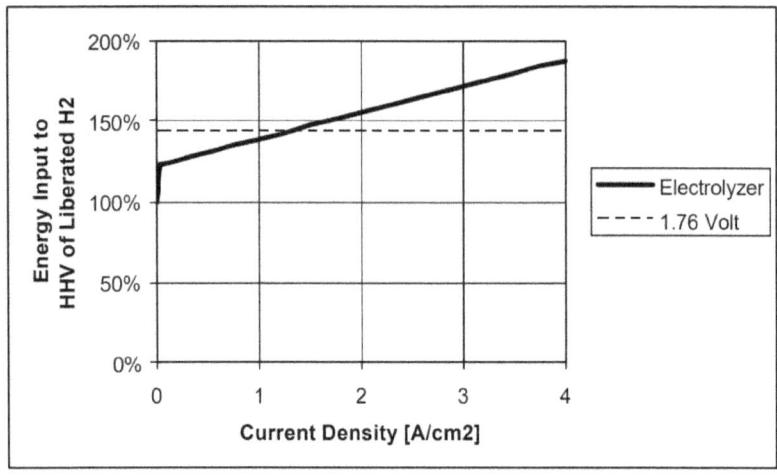

Figure 4   Energy input to electrolyze water compared to HHV energy of liberated hydrogen.

3.2 Reforming

Hydrogen can also be extracted from hydrocarbons by reforming. This chemical process is, in principle, an energy transfer process. The HHV energy contained in the original substance can be transferred to the HHV energy of hydrogen. Theoretically, no external energy is needed to convert a hydrogen-rich energy carrier like methane (CH_4) or methanol (CH_3OH) into hydrogen by autothermal steam reforming.

But in reality, thermal losses cannot be avoided and the HHV energy content of the original hydrocarbon fuel always exceeds the HHV energy contained in the generated hydrogen. The efficiency of hydrogen production by reforming is about 90%. Consequently, more CO_2 is released by this "detour" process than by direct use of the hydrocarbon precursors. But no obvious advantages can be derived concerning well-to-wheel efficiency and overall CO_2 emissions.

For most practical applications natural gas can do what hydrogen also does. There is no need for a conversion of natural gas into hydrogen which, as shown in this study, is more difficult to package and distribute than the natural energy carrier. The source energy (electricity or hydrocarbons) could be used directly by the consumer at comparable or even higher source-to-service efficiency and lower overall CO_2 emission. Upgrading electricity or natural gas to hydrogen does not provide a universal solution to the energy future, although some sectors of the energy market may prefer hydrogen. Fleet operation of vehicles may be one such application.

At today's energy prices, it is considerably more expensive to produce hydrogen by water electrolysis than by reforming fossil fuels. According to [5] it costs around $5.60 for every GJ of hydrogen energy produced from natural gas, $10.30 per GJ from coal, and $20.10 per GJ to produce hydrogen by electrolysis of water.

4. PACKAGING OF HYDROGEN

4.1 Compression of Hydrogen

Energy is needed to compress gases. The compression work depends on the thermodynamic compression process. The ideal isothermal compression cannot be realized. The adiabatic compression equation is more closely describing the thermodynamic process for ideal gases. The compression work depends on the nature of the gas. This is illustrated by the comparison of hydrogen with helium and methane in Figure 5:

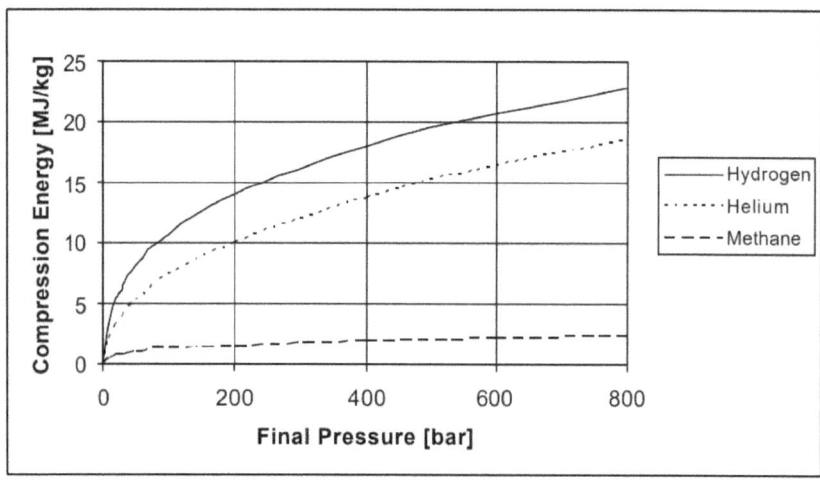

Figure 5   Adiabatic compression work for hydrogen, helium and methane

The energy consumed by adiabatic compression of monatomic Helium, diatomic hydrogen, and five-atomic methane from atmospheric conditions (1 bar = 100,000 Pa) to higher pressures is shown in Figure 2. Much more energy per kg is required to compress hydrogen than methane.

Isothermal compression follows a simpler equation:

$$W = p_0 V_0 \ln(p_1/p_0)$$

The same result is derived from the Nernst equation for the pressure electrolysis of water. In both cases, the compression work is the difference between the final and the initial energy state of the hydrogen gas.

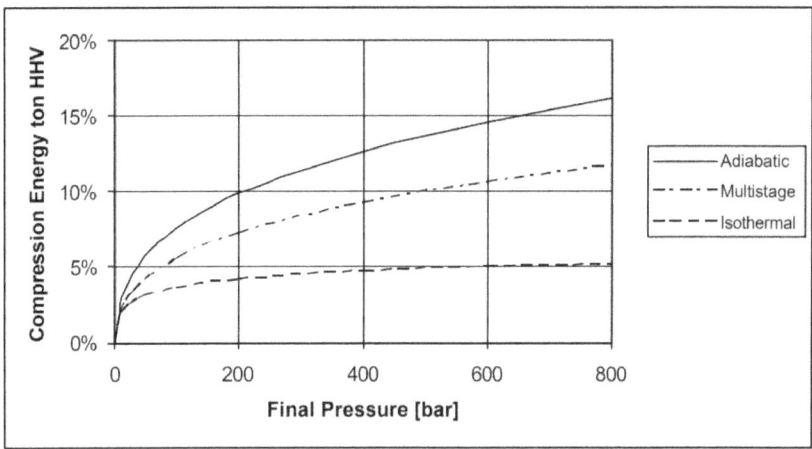

Figure 6   Energy required for the compression of hydrogen compared to its higher heating value HHV.

Figure 6 illustrates the difference between adiabatic and isothermal ideal-gas compression of hydrogen. Multistage compressors with intercoolers operate between these two limiting curves. Also, hydrogen readily passes compression heat to cooler walls, thereby approaching isothermal conditions. Numbers provided by a leading manufacturer [7] of hydrogen compressors show that the energy invested in the compression of hydrogen is about

7.2% of its higher heating value (HHV). This number relates to a 5-stage compression of 1,000 kg of hydrogen per hour from 1 to 200 bar. For a final pressure of 800 bar, the compression energy requirements would amount to about 13% of the energy content of hydrogen. This analysis does not include electrical losses in the power supply system.

4.2 Liquefaction of Hydrogen

Even more, energy is needed to compact hydrogen by liquefaction. Theoretically, only about 3.6 MJ/kg have to be removed to cool hydrogen down to 20K (-253°C) and another 0.46 MJ/kg to condense the gas under atmospheric pressure. About 4 MJ/kg are removed from room temperature hydrogen gas in the process, little compared to its energy content of 142 MJ/kg. But cryogenic refrigeration is a complex process involving Carnot cycles and physical effects (e.g. JouleThomsen) that do not obey the laws of heat engines. Nevertheless, the Carnot efficiency is used as a reference for the foregoing process analysis. For the refrigeration between room temperature (TR = 25°C = 298 K) and liquid hydrogen temperature (TL = -253°C = 20 K) one obtains a Carnot efficiency of or about 7%. The assumed single-step Carnot-type cooling process would consume at least 57 MJ/kg or 40% of the HHV energy content of hydrogen. This simple analysis does not include mechanical, thermal, flow-related, or electrical losses in the multi-stage refrigeration process. But by intelligent process design, the Carnot limitations may be partially removed. But the lower limit of energy consumption of a liquefaction plant does not drop much below 30% of the higher heating value of the liquefied hydrogen.

As a theoretical analysis of the complicated, multi-stage liquefaction processes is difficult, we present the energy consumption of existing hydrogen liquefaction plants.

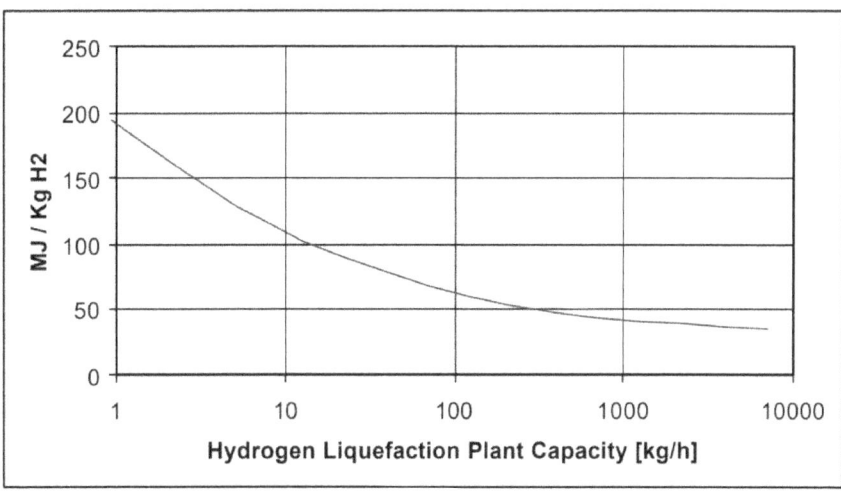

Figure 7   Typical energy requirements for the liquefaction of 1 kg hydrogen as a function of plant size and process optimization

The compilation reveals the following. Small (10 kg/h) liquefaction plants need about 100 MJ/kg, while large plants of 1000 kg/h or more capacity consume about 40 MJ of electrical energy for each kg of liquefied hydrogen. The actual liquefaction energy consumption for plants between 1 to 10,000 kg/h capacity is shown in Figure 7. The specific energy input decreases with plant size, but a minimum of about 40 MJ per kg H2 remains.

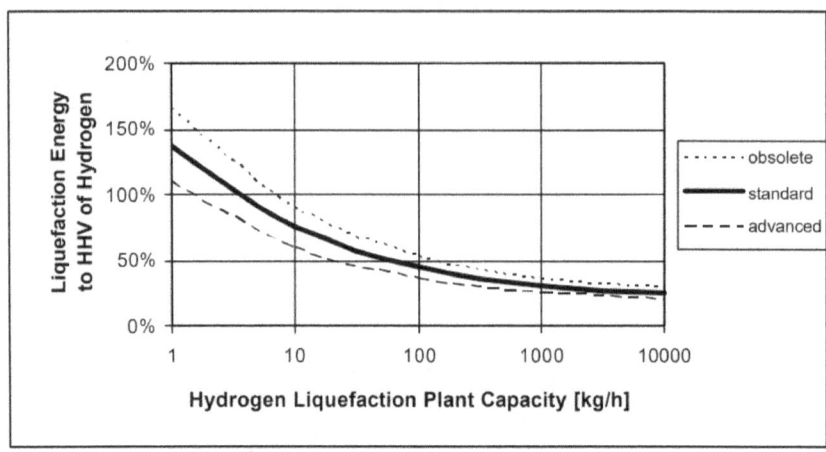

Figure 8   Actual energy requirement for the liquefaction of 1 kg hydrogen compared to HHV of hydrogen

In Figure 8 the required energy input is compared to the higher heating value HHV of hydrogen. For small liquefaction plants, the energy needed to liquefy hydrogen may exceed the HHV of the gas. But even with the largest plants (10,000 kg/h), at least 30% of the HHV energy is needed for the liquefaction process.

4.3 Physical Packaging of Hydrogen in Hydrides

At this time only a generalized assessment can be presented for the physical (e.g. adsorption on metal hydrides) storage of hydrogen in spongy matrices of special alloys like $LaNi_5$ or $ZrCr_2$. Hydrogen is stored by physical/chemical adsorption, i.e. by a very close, but not perfect bond between hydrogen atoms and the storage alloys. Heat is released when a hydrogen storage container is filled. The release of hydrogen at lower pressure is driven by an influx of heat proportional to the hydrogen liberation rate. According to [9] metal hydrides store only around 55-60 kgH_2/m^3 compared to 70 kgH_2/m^3 for liquid hydrogen. But 100 kg of hydrogen is contained in one cubic meter of methanol.

The energy balance shall be described in general terms. Again, energy is needed to produce and compress hydrogen. Some of this energy input is lost in form of waste heat. When hydrogen is released heat must be added. No additional heat is required for small liberation rates and containers designed for efficient heat exchange with the environment. Also, waste heat from the fuel cell may be used to heat the hydrogen storage cartridge.

One may wish to consider the transport energy for the heavy metal hydride cartridges. Not even two grams of hydrogen can be stored in a small 230 g metal hydride cartridge. This makes this type of hydrogen packaging impractical for automotive applications.

But the energy needed to package hydrogen in physical metal hydrides is more or less limited to the energy needed to produce and compress hydrogen to 30 bar pressure. The energy cost of hydrogen delivered to the customer in physical metal hydrides is thus lower than of compressed hydrogen gas delivered at 200 bar pressure.

4.4 Chemical Packaging of Hydrogen in Hydrides

Hydrogen may also be stored chemically in alkali metal hydrides. There are many options in the alkali group like LiH, NaH, KH, CaH2. But also complex binary hydride compounds like LiBH4, NaBH4, KBH4, LiAlH4, or NaAH4 are of interest and have been proposed as hydrogen sources. None of these compounds can be found in nature. All have to be synthesized from metals and hydrogen.

Let us consider the case of calcium hydride CaH2. The compound is produced by combining pure calcium metal with pure hydrogen at 480°C. Energy is needed to extract calcium from calcium carbonate (limestone) and hydrogen from water by the following endothermic processes

$$CaCO_3 \rightarrow Ca + CO_2 + 1/2\ O_2 \qquad + 808\ kJ/mol$$

$$H_2O \rightarrow H_2 + 1/2\ O_2 \qquad + 286\ kJ/mol$$

Some of the energy is recovered when the two elements are combined at 480°C by an exothermic process

$$Ca + H_2 \rightarrow CaH_2 \qquad -192\ kJ/mol$$

The three equations combine to the virtual net reaction

$$CaCO_3 + H_2O \rightarrow CaH_2 + CO_2 + O_2 \qquad + 902\ kJ/mol$$

Similarly, one obtains for the production of NaH and LiH from NaCl or LiCl

$$NaCl + 0.5\ H_2O \rightarrow NaH + Cl + 0.25\ O_2 \qquad + 500\ kJ/mol$$

and

$$LiCl + 0.5\ H_2O \rightarrow LiH + Cl + 0.25\ O_2 \qquad + 460\ kJ/mol$$

The material is then cooled under hydrogen to room temperature, granulated and packaged in airtight containers.

The hydrides react with water vividly under release of heat and hydrogen.

CaH2 + 2 H2O → Ca(OH)2 + 2 H2   - 224 kJ/mol

NaH + H2O → NaOH + H2   - 85 kJ/mol

LiH + H2O → LiOH + H2   - 111 kJ/mol

The reaction of hydrides with water produces twice the hydrogen contained in the hydride itself. Water is reduced while the hydride is oxidized to 16 hydroxides. The generated heat has to be removed by cooling and is lost in most cases. For the three representative hydrides, the energy balances are tabulated.

		Ca-Hydride	Na-Hydride	Li-Hydride
Hydride production from		CaCO$_3$	NaCl	LiCl
Energy to make hydride	kJ/mol	902	500	460
H$_2$ liberated from hydride	mol/mol	2	1	1
Production of H$_2$	g/mol	4	2	2
Energy input / H$_2$	kJ/g	225	250	230
=	MJ/kg	225	250	230
HHV of H$_2$	MJ/kg	142	142	142
Energy input / HHV of H$_2$	-	1.59	1.76	1.62

The results of this analysis are presented in Figure 9. The energy losses associated with the electrolytic decomposition of water, NaCl and LiCl have not even been considered.

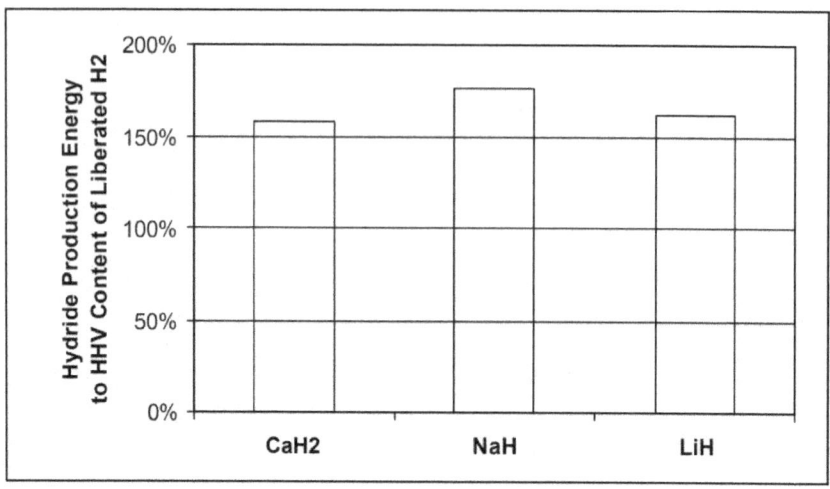

Figure 9   Energy needed to produce hydrides relative to HHV content of the liberated hydrogen

At least 160% of the HHV energy content of the liberated hydrogen has to be invested to produce the hydrides. The chemical packaging of hydrogen in alkali metal hydrides will therefore remain a solution for a limited number of practical applications. at least 60% of the input energy is lost in the process.

5. DELIVERY OF HYDROGEN

5.1 Road Delivery of Hydrogen

A hydrogen economy also involves hydrogen transport by trucks and ships. There are other options for hydrogen distribution, but road transport will always play a role, be it to serve remote locations or to provide backup fuel to filling stations at times of peak demand.

The comparative analysis is based on information obtained from the fuel and gas transport companies Messer-Griesheim, Esso (Schweiz), Jani GmbH , and Hover some of the leading providers of industrial gases in Germany and Switzerland. The following assumptions are made: Hydrogen (at 200 bar), liquid hydrogen, methanol, propane, and octane (representing gasoline) are trucked from the refinery or hydrogen plant to the consumer. Trucks with a gross weight of 40 tons (30 tons for liquid hydrogen) are fitted with suitable tanks or pressure vessels. Also, at full load, 40 kg of Diesel is consumed per 100 km. This is equivalent to 1 kg per ton per 100 km. The fuel consumption is reduced accordingly for the return run with emptied tanks. We assume the same engine efficiency for all transport vehicles.

While in most cases the transport is weight-limited, it is

limited by volume for liquid hydrogen as shown by the following sample. The useful volume of a large moving van, a box 2.4 m wide, 2.5 m high, and 10 m long, is 60 m3. But only 4.2 tons of liquid hydrogen can be filled into this box because the density of the cold liquid is only 70 kg/m3 or slightly more than that of heavy-duty Styrofoam. But space is needed for the container, thermal insulation, equipment, etc. There is room for only about 2.1 tons of liquid hydrogen on a large-size truck. This makes tracking of liquid hydrogen expensive because, despite its small payload, the vehicle has to be financed, maintained, registered, insured, and driven as any truck by an experienced driver. For the analysis, we assume the gross weight of the liquid hydrogen carrier is only 30 tons.

Furthermore, hydrogen pressure tanks can be emptied only from 200 bar to about 42 bar to accommodate for the 40 bar pressure systems of the receiver. Such pressure cascades are standard praxis today. Otherwise, compressors must be used to empty the content of the delivery tank into a higher-pressure storage vessel. This would not only make the gas transfer more difficult but also require additional compression energy as discussed below. As a consequence, pressurized gas carriers deliver only 80% of their freight, while 20% of the load remains in the tanks and is returned to the gas plant.

Each 40-ton truck is designed to carry a maximum of fuel. For methanol and octane, the tare load is about 26 tons, for propane about 20 tons. At 200-bar 18 pressure a 40-ton truck can carry 4 tons, but deliver only 3.2 tons of methane. Today, at 200 a pressure only 320 kg of hydrogen can be carried and only 288 kg are delivered by a 40-ton truck. This is a direct consequence of the low density of hydrogen, as well as the weight of the pressure vessels and safety armatures. In anticipation of technical developments, the analysis was performed for 4000 kg methane and 500 kg of hydrogen, of which 80% or 3200 kg and 400 kg, respectively, are delivered to the consumer. With this assumption, a deadweight of 39.6 tons has to be moved on the road to deliver 400 kg of hydrogen. On the return run, a heavy empty hydrogen truck consumes more diesel fuel than a much lighter empty gasoline carrier. The numbers in the following tables have been obtained for a 100 km delivery distance.

	Units	H2 Gas	Liquid H2	Methanol	Propane	Gasoline
Pressure	bar	200	1	1	5	1
Weight to customer	kg	40000	30000	40000	40000	40000
Weight from customer	kg	39600	27900	14000	20000	14000
Delivered weight	kg	400	2100	26000	20000	26000
HHV of fuel	MJ/kg	141.9	141.9	23.3	50.4	48.1
HHV energy per truck	GJ	57	298	580	1007	1252
Relative to gasoline	-	0.045	0.238	0.464	0.805	1
Diesel consumed	kg	79.6	57.9	54	60	54
Diesel HHV energy	GJ	3.56	2.59	2.41	2.68	2.41
Energy consumed to HHV energy delivered	%	6.27	0.87	0.42	0.27	0.19
Relative to gasoline	-	32.5	4.5	2.2	1.4	1
H2-efficiency factor	-	0.7	0.7	1	1	1
HHV energy delivered	GJ/d	876	876	1252	1252	1252
No. of trucks for same no. of serviced cars	-	15.4	2.9	2.2	1.24	1

The results of this analysis are presented in Figure 10. The energy needed to transport any of the three liquid fuels is reasonably small. It remains below 3% of the HHV energy content of the delivered commodity for a one-way delivery distance of 500 km.

But at almost any distance the relative energy consumption associated with the delivery of pressurized hydrogen becomes unacceptable. About 32 times more diesel fuel is required to deliver in the form of gaseous hydrogen compared to liquid gasoline. This factor is only about 4.5 for liquid hydrogen, but recall how much energy is required to liquefy the carried energy initially.

In our analysis, we do not consider improvements in the

fuel economy of both conventional engine and fuel cell vehicles. Today, the fuel economy of modern, clean Diesel engines is excellent but does not quite reach the HHV fuel economy of fuel cell vehicles. In both cases, the economy can be significantly improved by hybrid systems, mainly due to regenerative braking. But from well to wheel either fuel path leads to similar results concerning energy and CO_2 emissions. As 19 both technology offer potentials for improvements, no distinctive answer can be given at this time

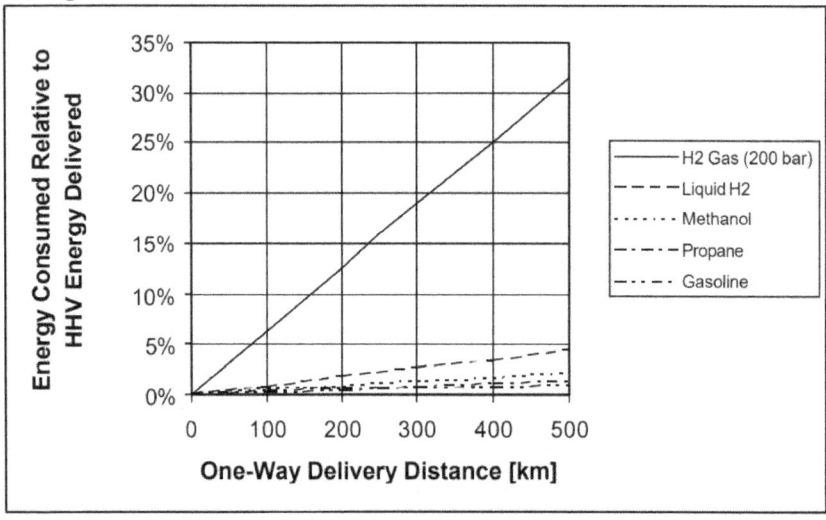

Figure 10   Energy needed for the road delivery of fuels compared to their HHV energy content

The following note may serve to illustrate the consequences of the scenario. A mid-size filling station on any major freeway easily sells 26 tons of gasoline each day. This fuel can be delivered by one 40-ton gasoline truck. Because of a potentially superior tank-to-wheel efficiency of fuel cell vehicles, we assume that hydrogen-fuelled vehicles need only 70% of the energy consumed by gasoline or Diesel vehicles to travel the same distance. Still, it would take 15 trucks to deliver compressed hydrogen (200 bar) energy to the station for the same daily amount of transport services, i.e. to provide fuel for the same number of passenger or cargo miles per day. Also, the transfer of pressurized hydrogen from those 15 trucks to the filling station takes much more time than draining gasoline from a single tanker into an underground storage tank. For safety reasons, the hydrogen filling station may have to close down for some hours every day.

Today about one in 100 trucks is a gasoline or diesel tanker. For surface transportation of hydrogen, one may see 115 trucks on the road, 15 or 13% of them transporting hydrogen. One out of seven accidents involving trucks would involve a hydrogen truck. Every seventh truck-truck collision would occur between two hydrogen carriers. This scenario is certainly unacceptable for many reasons.

5.2 Pipeline Delivery of Hydrogen

Hydrogen pipelines exist, but they are used to transport a chemical commodity from one to another production site. The energy required to move the gas has little is irrelevant, because energy consumption is part of the production costs. This is not so for hydrogen energy transport through pipelines. Normally, pumps are installed at regular intervals to keep the gas moving. These pumps are energized by energy taken from the delivery stream. About 0.3% of the natural gas is used every 150 km to energize a compressor to move the gas [14].

The assessment of the energy consumed to pump hydrogen through pipelines is derived from this natural gas pipeline operating experience. The comparison is done for equal energy flows. The same amount of energy is delivered to the customer through the same pipeline either contained in natural gas or hydrogen. In reality, existing pipelines cannot be used for hydrogen, because of diffusion losses, brittleness of materials and seals, incompatibility of pump lubrication with hydrogen, and other technical issues. The comparison further considers the different viscosities of hydrogen and methane.

Since the pumps run continuously, the power ratio also represents the ratio of the energy consumption for pumping.

Because of the low volumetric energy density of hydrogen, the flow velocity must be increased by over three times. Consequently, the flow resistance is increased significantly, but the effect is partially compensated for by the lower viscosity of hydrogen. Still, for the same energy flow about 4.6 times more energy is needed to move hydrogen through the pipeline compared to natural gas. As this energy is taken from the gas stream, more gas is fed into the pipeline than is delivered at the far end of the tube.

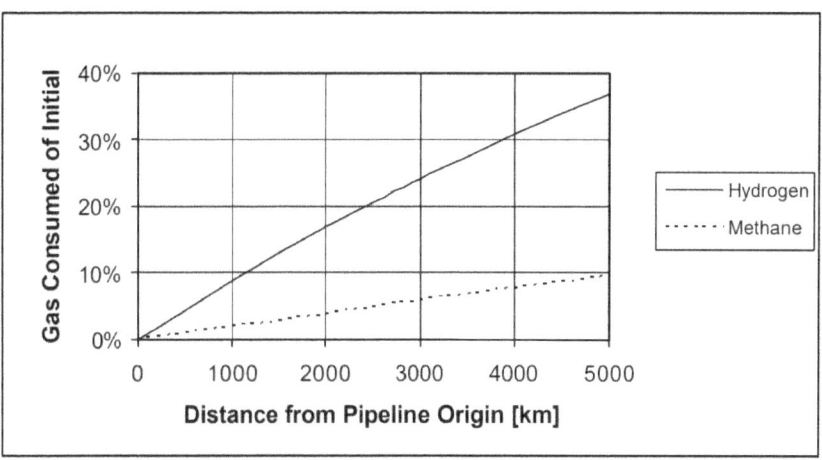

Figure 11   The fraction of the gas consumed to energize the pumps corresponds to the relative energy consumption (ratio of energy needed to HHV energy content) of the transported gases

Figure 11 shows the results of this approximate analysis. While the energy consumption for methane (representing natural gas) appears reasonable, the energy needed to move hydrogen through pipelines makes this type of hydrogen distribution difficult. Not 0.3% but at least 1.4% of the hydrogen flow is consumed every 150 km to energize the compressors. Only 60 to 70% of the hydrogen fed into a pipeline in Northern Africa would arrive in Europe.

5.3 Onsite Generation of Hydrogen

One option for providing clean hydrogen at filling stations and dispersed depots is the on-site generation of the gas by electrolysis. Again, the energy needed to generate and compress hydrogen by this scheme is compared to the HHV energy content of the hydrogen delivered to local customers. Natural gas reforming is not considered for reasons stated earlier.

The analysis is done for a single gas station serving 100

to 2,000 conventional road vehicles per day. On average, each car or truck is assumed to accept 60 liters (= 50 kg) of gasoline or diesel. For the 100 and 2000 vehicles per day, the energy equivalent would be about 1,700 to 34,000 kg of hydrogen per day, respectively. But on a tank-to-wheel basis, fuel cell vehicles consume less energy per driven distance than cars equipped with IC engines. Based on the HHV of both gasoline and hydrogen, we assume that fuel cell vehicles need only 70% of the energy consumed by IC engine vehicles to travel the same distance.

The key assumptions for continuous operation of the on-site hydrogen plant and the most important results are the following:

Vehicles / day	1/d	100	500	1000	1500	2000
Gasoline, Diesel / vehicle	kg	50	50	50	50	50
Fossil energy supplied	GJ/d	241	1,203	2,407	3,610	4,814
Efficiency factor	%	70	70	70	70	70
Hydrogen energy supplied	GJ/d	176	878	1,755	2,633	3,510
Hydrogen mass supplied	kg/d	1,188	5,938	11,877	17,815	23,753
Electrolyzer efficiency	%	70	75	78	79	80
AC/DC conversion	%	93	94	95	96	96
Energy for electrolysis	GJ/d	3259	1,195	2,274	3,332	4,388
Water needed	m³/d	11	53	107	160	214
Energy for water supply	GJ/d	8	36	68	100	132
H_2-compression, 200 bar	GJ/d	25	109	204	295	384
Total energy needed	GJ/d	292	1,340	2,546	3,727	4,903
Continuous power needed	MW	3	16	29	43	57
Relative to supplied H_2 HHV	%	173	159	151	147	146
Energy wasted per H_2 HHV	%	73	59	51	47	46

The electrolyzer efficiency varies with size from 70 to 80% for 100 and 2,000 vehicles per day, respectively. Also, losses occur in the AC-DC power conversion. Between 3 and 51 MW of power is needed for making hydrogen by electrolysis. Additional power is needed for the water make-up (0.09 to 1.52 MW) and the compression of the hydrogen to 200 bar (0.29 to 4.45 MW). In all, between 3 and 57 MW of electric power must be supplied to the station to generate hydrogen for 100 to 2,000 vehicles per day.

It may be of interest that between 11 and 214 m3 of water are consumed daily. The higher number corrresponds to about 2.5 liters per second.

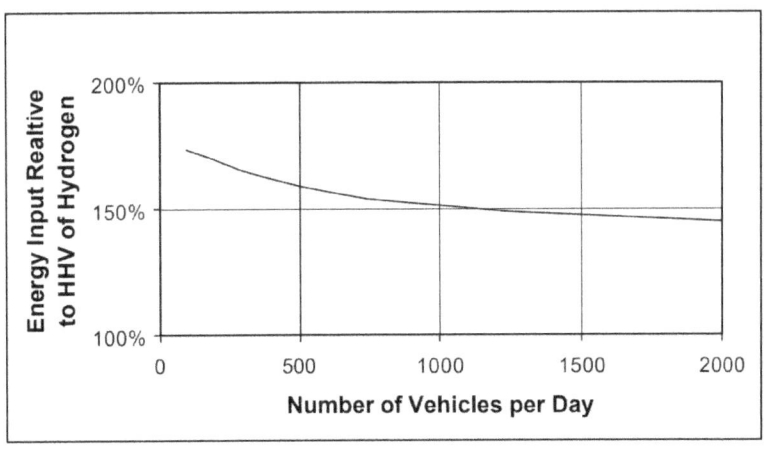

Figure 12   Energy needed for onsite generation of hydrogen by electrolysis and for compression to 200 bars at filling stations compared to the HHV energy content of the hydrogen delivered to road vehicles

The results of this analysis are presented in Figure 12. The total energy needed to generate and compress hydrogen at filling stations exceeds the HHV energy of the delivered hydrogen by 50%. The availability of electricity may certainly be questioned. Today, about one-sixth of the energy for end-use is supplied by copper wires. The generation of hydrogen at filling stations would require a 3 to 5 fold increase of the electric power generating capacity. The energy output of a 1 GW nuclear power plant is needed to serve twenty to thirty hydrogen filling stations on frequented highways.

6. TRANSFER OF HYDROGEN

Liquid can be drained from a full into an empty container by the action of gravity. There is no energy required unless the liquids are transferred from a lower to a higher tank, under controlled flow rates or accelerated conditions.

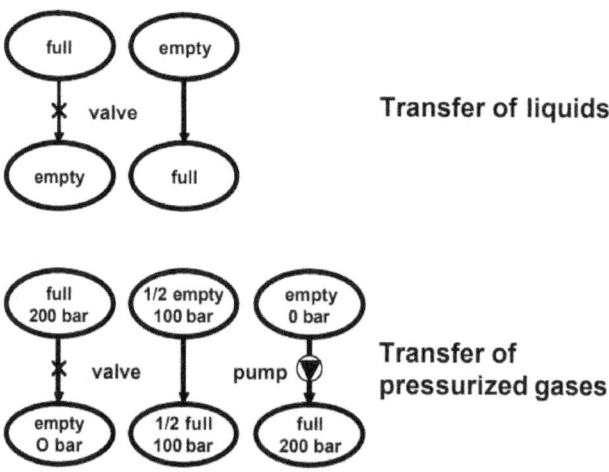

Figure 13   Schematic representation of the transfer of liquids and gases

The transfer of pressurized gases obeys different laws. Figure 13 may illustrate the point. Assume two tanks of equal volume, one full at 200 bar and the other empty at 0 bar pressure. After opening the valve between the vessels gas will flow into the empty tank, but the flow will cease when pressure equilibration is accomplished. Both tanks are half full or half empty. A pump is required to transfer the remaining content of the supply tank into the receiving tank. The transfer process may be complicated by temperature effects. The content of the full tank is cooled by the expansion process. At equal pressures, the density of the remaining gas is higher than that of the transferred gas in the other tank. As a consequence, more mass remains in the original vessel than is transferred into the empty one. Equal mass transfer is accomplished only after the temperatures have reached equilibrium after some time.

For the sample case considered, and for an ideal isothermal compression, the amount of energy required to complete the gas transfer by pumping is given by the difference of the total compression energy contained in the gas at the final pressure p2 and the intermediate pressure p1. The product p V (= R T) is the same for both compression processes.

$W = p_0 V_0 \ln(p_2/p_0) - p_0 V_0 \ln(p_1/p_0)$			
with W	[J/kg]	specific compression work	
p_0	[Pa]	initial pressure	
p_1	[Pa]	intermediate pressure	
p_2	[Pa]	final pressure	
V_0	[m³/kg]	initial specific volume	
For the sample case			
p_0	= 1 bar	= 1.0 x	10^5 Pa
p_1	= 100 bar	= 1.0 x	10^7 Pa

P_2	= 200 bar	= 2.0 x	10^7 Pa
V_0	= 11.11 m³/kg		
$P_0 V_0$	= 1.111 GJ/kg		

one obtains for the energy needed to transfer the remaining hydrogen from the half empty supply tank into the receiving tank by an isothermal compression

$$W = 0.77 \text{ GJ/kg}$$

or about 0.5% of the HHV energy content of the compressed hydrogen. For a more realistic adiabatic compression and including mechanical and electrical losses one would have obtained about 1%.

This number depends on the actual transfer conditions. Much more energy is needed to transfer hydrogen from a large 100 bar tank into a small container at 500 bar pressure. But it takes no additional energy to fill a small tank from a high-pressure vessel of substantial size. For automotive application, one aims at high-pressure tanks in vehicles and, as a consequence, has to use energy to transfer the hydrogen from large storage containers which cannot be subjected to high internal pressures. In any event, the transfer of hydrogen may add to the energy needs of a hydrogen economy.

7. SUMMARY OF RESULTS

The reported results are by no means final. The readers of this study are invited to refine the analysis and to contribute further details. The energy cost of producing, packaging, distributing, storing, and transferring hydrogen must have been analyzed in different contexts. The results of those studies may be used to verify, correct, or reject our numbers. Whatever the intent of this compilation is to create awareness about the weaknesses of a pure hydrogen economy. We are surprised to discover that the energy needed to run a hydrogen economy has never been fully assessed before.

Again, we would like to emphasize that the conversion of natural gas into hydrogen cannot be the solution to the future. Hydrogen produced by natural gas reforming may cost less than hydrogen obtained by electrolysis, but natural gas itself is as good as hydrogen or even better for many applications. Forgiven energy demand the well-to-wheel efficiency is reduced and, as a consequence, the emission of CO_2 is increased when natural gas is converted to hydrogen for daily use. For the final discussion, the key results are tabulated below.

	Energy cost in HHV of H_2	Factor	Path A gas	Path B liquid	Path C onsite	Path D hydride
Production of H_2						
Electrolysis	43%	1.43	1.43	1.43		1.22*
Onsite production	65%	1.65			1.65	
Packaging						
Compression 200 bar	8%	1.08	1.08			
Compression 800 bar	13%	1.13				
Liquefaction	40%	1.40		1.40		
Chemical hydrides	60%	1.60				1.60
Distribution						
Road, 200 bar H_2, 100 km	6%	1.06	1.06			
Road, liquid H_2, 100 km	1%	1.01		1.01		
Pipeline, 1,000 km	10%	1.10				
Storage						
Liquid H_2, 10 days	guess: 5%	1.05		1.05		
Transfer						
200 bar to 200 bar	1%	1.01	1.01		1.01	
Delivered to User						
Energy Input to HHV of H_2			1.65	2.12	1.66	1.95

* Only 50% of the liberated hydrogen comes from electrolysis

Four typical energy paths have been considered to interpret the results. These are:

A - Hydrogen is produced by electrolysis, compressed to 200 bar and distributed by road to filling stations or consumers

B - Hydrogen is produced by electrolysis, liquefied and distributed by road to filling stations or consumers

C- Hydrogen is produced onsite at filling stations or consumers

D- Hydrogen is produced by electrolysis and used to make alkali metal hydrides.

The analysis for ideal processes reveals that considerable amounts of energy are lost between the electrical source energy and the HHV hydrogen energy delivered to the consumer. For onsite hydrogen production, path C, the electrical energy input exceeds the HHV energy of the delivered hydrogen by a factor of at least 1.65. In the case of liquid hydrogen, path B, the factor is at least 2.12. For all stationary applications, the distribution of energy by copper wire will be a better choice than the use of hydrogen as an energy carrier.

But the problems of road delivery of compressed hydrogen have been discussed. It is unlikely that Path A can be realized. A better option would be the hydrogen distribution by short pipelines. To deliver hydrogen by chemical hydrides may provide practical solutions in some niche markets, but path D cannot become an important energy vector in a future economy.

Today, about 12% of the original fossil energy is lost between oil wells and filling stations for transportation, refining, and distribution. In a pure hydrogen economy, the losses would be considerably higher. If hydrogen could be chemically packaged in a synthetic liquid fuel, the overall energy consumption would be considerably lower.

7.1 The Limits of a Pure Hydrogen Economy

The results of this analysis indicate the weakness of a "Pure-Hydrogen-Only Economy" as depicted in Figure 14. Hydrogen is not only obtained by electrolysis, but also by chemical conversion of biomass. The economy is based on the natural H2O cycle, but the natural CO2 cycle is truncated and not fully used.

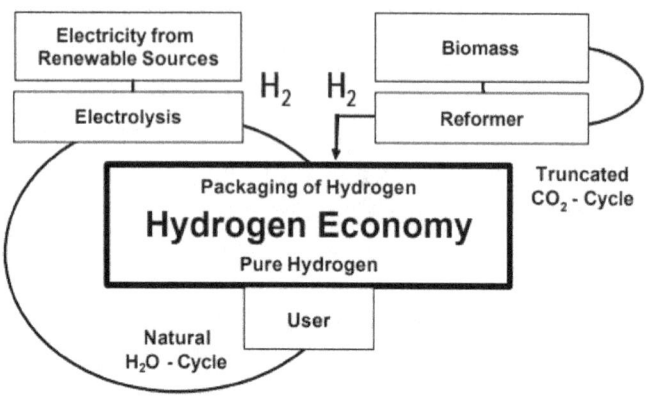

Figure 14   Pure Hydrogen Economy based on the natural cycle of water. Pure hydrogen is provided to the user

All difficulties with the pure Hydrogen Economy appear to be directly related to the nature of hydrogen. Most of the problems cannot be solved by additional research and development. We have to accept that hydrogen is the lightest of all gases and, as a consequence, that its physical properties do not fully match the requirements of the energy market. Production, packaging, storage, transfer and delivery of the gas, in essence all key component of an economy, are so energy consuming that alternatives should and will be considered. Mankind cannot afford to waste energy for idealistic goals, but economy will look for practical solutions and select the most energy-saving procedures. The "Pure-Hydrogen-Only-Solution" may never become reality.

The degree of energy waste certainly depends on the chosen path. Hydrogen generated from rooftop solar electricity and stored at low pressure in stationary tanks may be a viable solution for private buildings. On the other hand, hydrogen 29 generated in the Sahara desert, pumped to the Mediterranean Sea through pipelines, then liquefied for sea transport, docked in London, and locally distributed by trucks may not provide an acceptable energy solution at all. Too much energy is lost in the process to justify the scheme. But there are solutions between these two extremes, niche applications, special cases, or luxury installations. This study provides some clues for the strengths and weaknesses of the energy carrier hydrogen.

As stated in the beginning, hydrogen may be the only link between physical energy from renewable sources and chemical energy. It is also the ideal fuel for modern clean energy conversion devices like fuel cells or even hydrogen engines. But hydrogen is not the ideal medium to carry energy from primary sources to distant end-users. New solutions must be considered for the commercial bridge between electrolyzer and fuel cell.

7.2 A Liquid Hydrocarbon Economy

The ideal energy carrier is a liquid with a boiling point above 80°C and a solidification point below -40°C. Such energy carriers stay liquid under normal climate conditions and at high altitudes. Gasoline, diesel, and methanol are good examples of such fuels. They are in common use not only because they can be extracted from crude oil, but mainly because they qualify for widespread use because of their physical properties.

Oil companies convert crude oil into gasoline and diesel fuels. Even if oil had never been discovered, the world would not use synthetic hydrogen, but one or more synthetic hydrocarbon fuel. Gasoline, diesel, heating oil, etc. have emerged as the best solutions to handling, storage, transport, and energetic use. With high certainty, such liquids will also be synthesized from hydrogen and carbon in a distant energy future. Fortunately, methanol and ethanol can also be derived from plants by biological fermentation processes.

There are several synthetic hydrocarbons to be considered. One of the prime choices may be methanol. It carries four hydrogen atoms per carbon atom. It is liquid under normal conditions. The infrastructure for liquid fuels exists. Also, methanol can either be directly converted to electricity by Direct Methanol Fuel Cells (DMFC), Molten Carbonate Fuel Cells (MCFC), and Solid Oxide Fuel Cells (SOFC). It can also be reformed easily to hydrogen for use in Polymer Electrolyte Fuel Cells (PEFC or PEM). Methanol could become a universal fuel for fuel cells and many other applications.

Liquid Hydrocarbon Economy

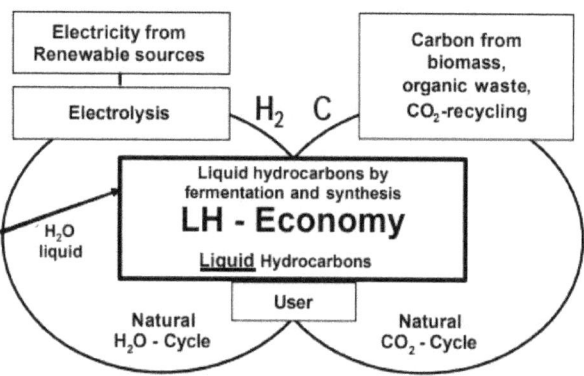

Figure 15   A Liquid Hydrogen Economy is based on the two natural cycles of water and carbon dioxide. Natural and synthetic liquid hydrocarbons are provided to the user

Figure 15 shows a schematic of a "Liquid Hydrocarbon Economy" (in short: "LH Economy"). It is based on the two natural cycles of water and carbon dioxide. Carbon from the biosphere may become the key element in a sustainable energy future. It could come from biomass, organic waste, and captured CO_2. Typically, biomass has a hydrogen-to-carbon ratio of two. In the methanol synthesis, two additional hydrogen atoms are attached to every bio-carbon. Instead of converting biomass into hydrogen, hydrogen from renewable sources or even water could be added to biomass to form methanol by a chemical process. In an LH economy, carbon atoms will stay bound in the energy carrier until their final use. They are then returned to the atmosphere (or recycled). This is true not only for methanol but also for ethanol or other synthetic hydrocarbons. The suggested scheme should be seriously considered for the planning of a clean and sustainable energy future.

7.3 Liquid Hydrocarbons

Any synthetic liquid fuel must satisfy several requirements. It should be liquid under normal pressure at temperatures between -40°C and 80°C, be nontoxic, be useful for IC engines, easy to synthesize, etc. The chemicals tabulated below satisfy the liquidity criteria. They may serve to illustrate that several options exist for the synthesis of liquid hydrocarbons from hydrogen and carbon. But aspects of manufacturing, safety, combustion, etc., all well-known to the experts, will eliminate some or add new options to the list.

The following liquid hydrocarbons are considered:

A	Methanol	CH_4O	or CH_3OH
B	Ethanol	C_2H_6O	or CH_3CH_2OH
C	Dimethlyether (DME)	C_2H_6O	or CH_3OCH_3
D	Ethylmethylether	$C_4H_{10}O$	or $CH_3OC_2H_5$
E	2-Methylpropane (Isubutane)	C_4H_{10}	or $CH_3CH(CH_3)CH_3$
F	2-Methylbutane (Isopentane)	C_5H_{12}	or $CH_3CH(CH_3)CH_2CH_3$
G	Ethylbenzol	C_8H_{10}	or $C_6H_5CH_2CH_3$
H	Methylcyclohexane (Toluol)	C_7H_{14}	or $C_6H_5CH_3$
I	Octane	C_8H_{18}	or $CH_3(CH_2)_3CH_3$
J	Ammonia	NH_3	
K	Hydrogen (for comparison)	H_2	

Methanol, Ethanol, DME, Toluol and Ammonia, all having relatively simple molecular structures, may become the preferred synthetic energy carriers of the future in competition with liquid (or 800 bar) hydrogen. The ten substances are characterized by the following technical numbers:

Fuel	Mol. Weight mole	Density kg/m³	H_2-Content moleH$_2$/mole	H_2-Density kgH$_2$/m³	HHV MJ/kg	Energy per Volume GJ/m³
A	32	792	0.125	99	22.7	17.97
B	46	789	0.130	103	29.7	23.45
C	46	666	0.130	87	31.7	21.14
D	74	714	0.135	96	28.5	20.34
E	58	557	0.172	96	49.4	27.54
F	72	620	0.167	103	48.7	30.17
G	106	866	0.094	82	43.1	37.30
H	112	769	0.125	96	34.9	26.85
I	114	703	0.158	111	48.0	33.73
J	17	770	0.176	136	22.5	17.35
K	2	70	1.000	70	141.9	9.93

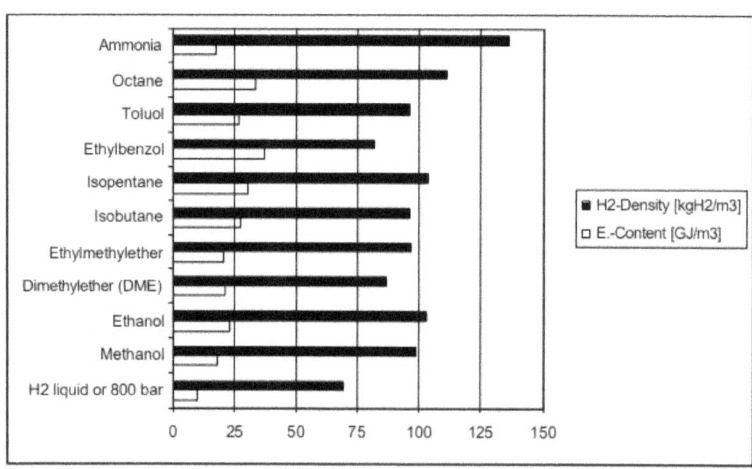

Figure 16   Hydrogen density and HHV energy content of selected synthetic liquid hydrocarbon fuels and Ammonia

The results are depicted in Figure 16. Any one of the nine hydrocarbon fuels contains more hydrogen per cubic meter than is contained in the same volume of liquefied or 800 bar compressed hydrogen. Ammonia even contains even 136 kg of hydrogen per cubic meter. Also, the energy carried by the hydrocarbons is between two and almost four times greater than the energy contained in the same volume of liquid hydrogen. If one wants to distribute hydrogen, obviously the best way is combining it with carbon to liquid fuel. It may be of interest to observe that the gasoline-like Octane seems to be the best hydrogen carrier and also ranks among the best-concerning energy content per volume. The synthesis of Octane from bio-carbon and water may pose an attractive solution for an energy economy based on renewable energy sources and the recycling of carbon dioxide.

8. CONCLUSIONS

Time has come to shift the attention of energy strategy planning, research, and development from a "Hydrogen Economy" to a "Synthetic Liquid Hydrocarbon Economy" and to direct manpower and resources to find technical solutions for a sustainable energy future which is built on the two closed clean natural cycles of water and CO2 or hydrogen and carbon. If carbon is taken from the biosphere or recycled from power plants ("bio-carbon") and not from fossil resources ("carbon"), the "Synthetic Liquid Hydrocarbon Economy" will be environmentally as benign as a "Pure Hydrogen Economy".

BOOKS BY THIS AUTHOR

Solid State Battery: A Battery From Future

A solid-state battery is a battery technology that uses solid electrodes and a solid electrolyte, instead of the liquid or polymer gel electrolytes found in lithium-ion or lithium polymer batteries. A solid-state battery is a battery technology that uses solid electrodes and a solid electrolyte, instead of the liquid or polymer gel electrolytes found in lithium-ion or lithium polymer batteries.

Hybrid Electric Vehicles: Vehicle Of Future

A hybrid electric vehicle (HEV) is a type of hybrid vehicle that combines a conventional internal combustion engine (ICE) system with an electric propulsion system (hybrid vehicle drive train). The presence of the electric power train is intended to achieve either better fuel economy than a conventional vehicle or better performance. There is a variety of HEV types and the degree to which each function as an electric vehicle (EV) also varies. The most common form of HEV is the hybrid electric car, although hybrid electric trucks (pickups and tractors) and buses also exist.

Internet Of Things-Iot : Definition, Characteristics, Architecture, Enabling Technologies, Application & Future Challenges

The Internet of things refers to a type of network to connect anything with the Internet-based on stipulated protocols through information sensing equipment to conduct information exchange and communications to achieve smart recognitions, positioning, tracking, monitoring, and administration. In this paper, we briefly discussed what IoT is, how IoT enables different technologies, its architecture, characteristics & applications, IoT functional view & what are the future challenges for IoT.

www.ingramcontent.com/pod-product-compliance
Lightning Source LLC
Chambersburg PA
CBHW050314220526
45465CB00005B/1979